PowerKids Readers:

ROAD MACHINES™

Road Pavers

Joanne Randolph

The Rosen Publishing Group's
PowerKids Press™
New York

For Joseph Hobson, with love

Published in 2002 by The Rosen Publishing Group, Inc.
29 East 21st Street, New York, NY 10010

First Edition

Book Design: Michael Donnellan

Photo Credits: all photos by Naima Rauam/artpm.com.

Randolph, Joanne.
  Road pavers / Joanne Randolph.
    p. cm. — (Road machines)
  Includes bibliographical references and index.
  ISBN 0-8239-6040-4 (library binding)
  1. Asphalt pavers—Juvenile literature. [1. Road machinery.] I. Title
TE273 .R36 2002
625.8'5—dc21

                                                    2001000655

Manufactured in the United States of America

Contents

This is a paver.

4

5

Pavers pave the road. They spread the hard, top layer of tar that you walk and drive on.

Pavers have a hopper on the front. The hopper is like a big, square box.

9

Dump trucks put hot, soft tar into the hopper.

Pavers have spreaders at the back of the truck. These spreaders are called screeds.

13

The screed helps to spread the tar in a thin layer on the road.

15

Men fill in any holes left by the paver. These men shovel tar from the back of the hopper. Then they put the tar inside the holes in the road.

Another person needs to drive the paver. He stands on top and drives it slowly. The other men check to make sure the paver is doing a good job.

19

It takes a big team to pave a road.

Words to Know

hopper

paver

screed

Here is another book to read about pavers:

Diggers and other Construction Machines
by Jon Richards
Copper Beech Books

To learn more about pavers, check out this
Web site:
www.field-guides.com/html/pavers.html

23

Index

Word Count: 142

Note to Librarians, Teachers, and Parents

PowerKids Readers are specially designed to help emergent and beginning readers build their skills in reading for information. Simple vocabulary and concepts are paired with photographs of real kids in real-life situations or stunning, detailed images from the natural world around them. Readers will respond to written language by linking meaning with their own everyday experiences and observations. Sentences are short and simple, employing a basic vocabulary of sight words, as well as new words that describe objects or processes that take place in the natural world. Large type, clean design, and photographs corresponding directly to the text all help children to decipher meaning. Features such as a contents page, picture glossary, and index help children get the most out of PowerKids Readers. They also introduce children to the basic elements of a book, which they will encounter in their future reading experiences. Lists of related books and Web sites encourage kids to explore other sources and to continue the process of learning.